D 93

Diese Mitteilungen setzen eine von Erich Regener begründete Reihe fort, deren Hefte auf der vorletzten Seite genannt sind.

Das Max-Planck-Institut für Aeronomie vereinigt zwei Institute, das Institut für Stratosphärenphysik und das Institut für Ionosphärenphysik.

Ein (S) oder (I) beim Titel deutet an, aus welchem Institut die Arbeit stammt.

Anschrift der beiden Institute:

3411 Lindau

TIME PATTERN OF IONIZING RADIATION
IN BALLOON ALTITUDES IN HIGH LATITUDES

by

G. Pfotzer, A. Ehmert, and E. Keppler

ISBN 978-3-540-02880-2   ISBN 978-3-642-88499-3 (eBook)
DOI 10.1007/978-3-642-88499-3

Die vorliegende Arbeit wurde als Schlußbericht
eines Forschungs-Kontraktes mit dem

    OFFICE OF SCIENTIFIC RESEARCH, OAR, through the
    European Office of the Office of Aerospace
    Research, United States Air Force,

zusammengestellt.

Durch ihre Veröffentlichung in den "Mitteilungen
aus dem Max-Planck-Institut für Aeronomie" soll
sie auch einem größeren Kreis interessierter Fach-
kollegen zugänglich gemacht werden.

Da der volle Informationsgehalt der hier zusammen-
gestellten und von amerikanischen Gruppen gleichzeitig durchgeführten Messungen noch durch Aus-
tausch und gemeinsame Diskussion der wechselsei-
tigen Ergebnisse ausgeschöpft werden soll, wurde
aus praktischen Gründen die englische Fassung
beibehalten.

CONTENTS

I.	Introduction	1
II.	Review of the General Background	
	Galactic and Solar Cosmic Radiation	1
	Solar Protons	3
	Bremsstrahlung of Electrons	4
III.	Location of the Balloon Flights	6
IV.	Detectors	
	Single Counter	8
	Threefold coincidence telescope	9
	Efficiency	10
	Ionization chamber	10
	Temperature coefficient	12
	Energy threshold for detection of particles or photons	13
	Relations between the counting rates of the single counter and the telescope bearing on the directional distribution of any radiation and the range and energy spectrum of charged particles	13
	Conversion of the range distribution of protons into an energy spectrum	15
	Relation between the counting rate of the single counter, the pulse rate of the ionization chamber and the average energy of X-Rays	16
V.	Circuitry of the TESIO unit	16
VI.	Ground Station	20
VII.	Balloons and Accessories	21
VIII.	Indications for Launching	23
IX.	Results	25
	1. General Remarks	25
	2. Representation of Data	26

a.	High Altitude Records	27
b.	Riometer	29
c.	Magnetic Components	29

3. Comments on the various flights 30
The events around July 11th, 12th, and 13th 38
The events of July 18th and 20th 42
The event on July 28th 44

Acknowledgement 46
Literature 47

Zusammenfassung

Zwischen dem 12. Juni und dem 2. August 1961 wurde eine Serie von 23 Flügen mit unbemannten Ballonen zur Messung der Kosmischen Strahlung, Solarer Protonen und Röntgenstrahlungsausbrüchen in großen Höhen über der Nordlichtzone erfolgreich durchgeführt. Die automatisch registrierenden Geräte und die Besonderheiten der Meßmethode werden im einzelnen beschrieben. Die Meßergebnisse sind in 45 Kurvenblättern (Teil B) zusammengefaßt.

Neben den Röntgenstrahlungsausbrüchen, die verhältnismäßig häufig in der Nordlichtzone auftreten, wurden 4 Fälle von Injektionen solarer Protonen am 12., 18., 20. und 27. Juli registriert, während sich 2 sehr aktive Fackelgebiete über die Sonnenscheibe hinwegbewegten. Die verschiedenen Ereignisse werden kurz diskutiert. Die hier beschriebenen Messungen fanden zu einem großen Teil gleichzeitig mit solchen ähnlicher Art statt, die von amerikanischen Gruppen durchgeführt wurden.

I. Introduction

Until 20 years ago most of our knowledge on cosmical objects was based merely on the analysis of visible light reaching the earth from outer space. This changed when appropriate techniques had been developed to utilize increasingly the perception that every kind of radiation or - generally spoken - any structure of energy hitting the earth carries information on its origin and on the conditions in space which influenced its qualities and propagation. This holds especially also for radiation of charged particles.

In the main part of this paper we shall report on measurements of such charged particles and of high energy photons produced by them. These measurements were carried out by means of unmanned balloon flights for the most part launched in the auroral zone. A brief introduction in the general background of the subject, some details of the technical features and a compilation of the results will be given.

II. Review of the General Background

Galactic and Solar Cosmic Radiation. The nucleonic radiation in contrast to electromagnetic waves and electrons represents a continuous influx of matter. One part of it, the Galactic Cosmic Radiation (Galactic C. R.), originates very probably from extremely active stellar atmospheres or special objects like magnetic stars or supernovae. It is presumably injected in a large scale acceleration mechanism in the galaxy and finally trapped in very extended and relatively weak magnetic fields present in the so-called Galactic radio-corona which surrounds the proper galaxy. (L1) A minor fraction of high energy charged particles originates at the sun.

Since the early days of Cosmic Ray research there has always been considerable interest in the time variations of the particle flux and the aim has been to disclose a siderial periodicity accentuating spatially fixed directions which would point to limited sources of the radiation. So far, however, no persisting periodicity of this kind could be found in the limits of experimental errors and hence it must be concluded that the flux of Galactic C. R. outside the solar controlled region is essentially isotropic. One case of a possible point source with decreasing efficiency claimed by Sekido et al. (L2) does scarcely affect this general picture. It turned out that all variations of the charged particle flux with amplitudes normally in the order of percents but sometimes reaching a manifold of the average level are controlled by solar activity. There can be distinguished two kinds of variations:

1) A modulation of the Galactic C. R. by hydromagnetic fields in the interplanetary space

2) Increases which are due to the emission of high energy atomic nuclei like in the Galactic C. R. consisting also mainly of protons from active regions of the sun during the flashing up of solar flares (therefore normally called solar protons only for simplicity).

Although we do not know so far much about the modulation mechanism involved, there is no doubt that the variations of the Galactic C. R. indicate the build up and dilution of magnetized plasma clouds in the interplanetary space which are ejected from the sun.

When the average density of hydromagnetic plasma in the interplanetary space increases with increasing solar activity the level of the Galactic C. R. decreases. This holds as well in the average over the eleven year sunspot cycle as it is very spectacularly expressed during violent "earth storms". These are considered to be an immediate consequence of the ejection of plasma clouds from solar flare regions towards the earth.

When the plasma flows around the earth a characteristic decrease and gradual recovery of the radiation intensity, the so-called "Forbush decrease" (F.d.) is observed. These variations are related effectively to some kind of "interplanetary weather" in analogy to the variations of the barometer level with our normal weather.

Solar Protons. The emissions of solar protons with energies extending from several MeV sometimes up to several BeV are interesting in various respects. Referring to the modulation of the Galactic radiation we should like to emphasize firstly their significance as space probes.

Each proton event exhibits special features depending on the transient magnetic field regime. Rather detailed indications for field configurations are given by the delay of arrival of the particles and by the directions from where they arrive at the boundary of the magnetosphere. (L3, L4, L5) These features refer to the parts played by magnetic fields either as barriers or as guiding channels for the solar protons. It can also be considered as secured that sometimes protons ejected from a flare in a certain plage region are trapped in a magnetized plasma beam (magnetic bottle) originating from a preceding flare in the same plage region. (L6) A high flux of energetic particles is thus confined to a relative slowly expanding volume. A steep increase of ionizing radiation indicates when the magnetized plasma cloud reaches and eventually encloses the earth.

Besides this relatively recent aspect, the solar proton events have always engrossed considerable interest as examples for possible acceleration mechanisms transferring high energies to charged particles in stellar atmospheres. We scarcely need to emphasize that the sun is the only natural source of high energy particles which can be studied in full action and with associated phenomena like flashes of visible or ultra-violet light, radio-wave bursts and the ejection of plasma clouds. In most cases these phenomena can be linked unambiguously to the production of fast particles. This promises a gradual approach not only to the understanding of this special process but also to those on larger scale generating the bulk of the Galactic C. R.

So far we have stressed only the aspects which at present deserve mainly our scientific interest. There is, however, another one of vital importance for astronauts. It could be inferred that the radiation dosage due to solar protons in free space reaches sometimes very harmful levels. (L7, L8, L9) Though the various estimates are not yet well founded we cannot doubt that the dosage in a spaceship without special shielding would have exceeded the letal value by far during at least 3 events (February 23, 1956, May 12, 1959, November 12, 1960). This last aspect does not only justify - like the merely scientific ones - a thorough exploration of this subject with all available methods but demand it even straight on.

Bremsstrahlung of electrons. A sporadic flux of X-Rays in high altitudes has to be mentioned as a second kind of radiation in which we are interested. It was detected firstly by Van Allen and his co-workers in 1953 by means of the rockoon technique (rockets fired from balloons in altitudes of about 25 kms). (L10, L11, L12) These X-Rays are mainly confined to the auroral zone what implies an influence of the earth magnetic field and hence a coherence with primary charged particles. The latter could be identified as electrons by Davis et al. (L13) and McIlwain (L14) what confirmed the original deductions of the first mentioned authors. As we assume today these electrons are closely linked to the radiation belts which were discovered also by Van Allen and his associates in 1958 by means of the first American satellite Explorer I. (L15) A classification of the essential aspects of this phenomenon is given in a preceding paper. (L16) It shall here be summarized only briefly. There is a nearly continuous flux of low energy X-Rays ($E \approx 20$ keV) in altitudes exceeding 40 kms and there are violent bursts of X-Rays with energies between 20 and 100 keV lasting from minutes to several hours observable in balloon altitudes as low as 25 to 30 kms. (L17, L18, L19, L20, L21, L22) The more energetic bursts are normally well correlated with the variations of the magnetic field components and with the Cosmic noise absorption (CNA) as for example measured by the riometers (Relative Ionospheric Opacity meters) (L23). In other cases when smaller energies are involved no correlation with the magnetic disturbances is clearly expressed.

In general the energy of the primary electrons is not sufficient for penetrating the undisturbed magnetic field of the earth and for being detected in the region of the auroral zone. If they stem from outside of the magnetosphere they must be pushed in by hydromagnetic processes or otherwise be accelerated in the magnetosphere. In each case an intermediate step of trapping in the Van Allen belts might be involved. There are indications that the electrons were precipitated from the belts when strong magnetic disturbances occurred but no clear picture of the process could hitherto be derived.

The most promising way to approach to a better understanding of these phenomena is still the broadening of the empirical base by performing measurements of sufficient duration to show time patterns which can be correlated with other geophysical parameters. Since the satellites are in too high orbits for this purpose and the rockets do not stay long enough at constant altitudes, measurements of this kind are still in the domain of balloon work.

III. Location of the Balloon Flights

The preceding review of charged particle radiations has shown that they can be utilized as very effective space probes giving information on transient force fields in the interplanetary space on acceleration processes on the sun, and the dynamics in the magnetosphere of the earth.

At a first glance this very quality of the charged particles to be strongly affected by electromagnetic fields, seems to be disadvantageous for the interpretation of measurements on the earth because the particles are also strongly deflected by the earth magnetic field. Since, however, the latter acts as some kind of energy spectrometer additional information can be drawn out by performing simultaneous measurements at different locations. We proposed therefore to coordinate balloon flights in Europe and USA such that the measurements cover the same period. Fortunately we found the partnership of two American groups: one of the University of California, Berkeley, headed by Prof. R.R. Brown, the other of the University of Minnesota, headed by Prof. J.R. Winckler. Especially the cooperation with the latter group was successful in that it covered the very interesting period of enhanced solar activity between July 12th and July 27th, 1961.

Concerning the choice of locations for balloon flights the following considerations hold: The magnetic rigidities of charged particles which can arrive at the top of the atmosphere at a given geomagnetic latitude must exceed a corresponding cut off value. The latter decreases from the equator towards the poles. On the other hand the primary charged nuclei can manifest themselves on ground stations by a secondary nucleonic component if their kinetic energy exceeds approximately 500 MeV. (Atmospheric cut off for sea level) This corresponds to the geomagnetic cut off for a geomagnetic latitude of $\sim 60°$. It means that in principle up to $60°$ geom. latitude most of the major variations of the primary

C. R. are reflected by the variations of the nucleonic component measured on ground by neutron monitors. At geom. latitudes above 60° where protons with energies $<$ 500 MeV are also admitted geomagnetically at the top of the atmosphere the secondaries of the particles with lowest energies can no more penetrate to sea level to record all geomagnetically admitted particles in these regions, the measurements have to be carried out at increasingly higher altitudes.

Since the particles with lowest energies are the most sensitive space probes and because solar proton events with energies of the particles $<$ 500 MeV are much more frequent than those which can also be observed at sea level it is advisable to carry out the measurements at high latitudes and altitudes. This lead us to perform the flights at Kiruna in Northern Sweden where also the facilities of the Geophysical Observatory by courtesy of the Royal Swedish Academy of Sciences had been at our disposal. This station is situated near the low latitude border of the auroral zone. It must be emphasized that it is <u>practically ideal for balloon work</u> because not only the rather frequent low energy proton events but also the X-Ray bursts caused by electron precipitations can be investigated most effectively.

The same and even better conditions are of course prevailing at Fort Churchill where simultaneous or overlapping flights were performed by the University of Minnesota Group which also launched a large number of flights at Minneapolis. 4 flights of short duration took place at Lindau. The relevant data for these stations are compiled in table 1.

It is seen that for Kiruna at balloon altitudes (at an atmospheric depth of 10 mb) the geomagnetic cut off for undisturbed magnetic field conditions just equals the atmospheric cut off.

TABLE 1

Locations and geomagnetic cut off data for the stations where simultaneous balloon flights were performed in summer 1961. (Cut off data after Quenby and Wenk)(L24).

E_{pkin} = kinetic energy of protons corresponding to the cut off rigidity. R_p = range of primary protons with E_{pkin} just above the cut off in terms of pressure level in mb to which they can penetrate.

Station	geographic long.	lat.	rigidity [GV]	E_{pkin} [MeV]	R_p [mb] just above cut off	equiv. geom. lat.
Kiruna	20.4°E	67.9°N	0.49	128	10	64.8°N
Ft.Churchill	265.9°	58.8°	0.20	21	0.5	70° N
Minneapolis	266.7°	45.0°	1.38	826	-	56.5°N
Lindau	10 °	51.5°	2.93	2.41	-	48.3°N

IV. Detectors

The principal series of flights were performed with TESIO units which represent combinations of a GM-counter <u>tele</u>-scope, a <u>s</u>ingle GM-counter, and an <u>i</u>onization chamber. The single counter was mounted with vertical axis. This instrumentation allows to discriminate between charged particles and photons and to estimate an average energy of the photons.

<u>Single Counter</u>. The single counter and also those of the telescope were of the Victoreen type 1 B 85. Their efficiency for X-Rays of different energies is plotted in

Fig. 1 after a curve of relative efficiency given by the manufacturer and normalized to an efficiency of $\varepsilon_s = 0,5\ \%$ for 70 keV, a value given by Anderson (L18). The wall thickness of the counter amounts to 0,11 mm aluminium equivalent to an absorption layer of 30 mg/cm^2, the inner diameter to 1,9 cm and the effective length to 6,5 cm. The counting rate N of the counter can be expressed by the formula:

$$N_s = G_s(\mu) \cdot \varepsilon_s \cdot I_o \qquad (1.0)$$

whereby G_s is the geometrical factor for the single counter, given below, I_o is the particle- or quanta-flux from the vertical per unit area, unit solid angle and unit time. The flux $I(\vartheta)$ coming from a direction inclined by the angle ϑ towards the zenith shall be approximated as usual by

$$I(\vartheta) = I_o \cos^\mu \vartheta \qquad (2.0)$$

then we get for different μ

$$\mu = 0: \quad G(0) = \frac{\pi^2}{2}\ al\ (1 + 1/2\ a/l) \qquad (3.0)$$

$$\mu = 1: \quad G(1) = \frac{2\pi}{3}\ al\ (1 + \pi/4\ a/l) \qquad (3.1)$$

$$\mu = 2: \quad G(2) = \frac{\pi^2}{8}\ al\ (1 + a/l) \qquad (3.2)$$

$$\mu = 3: \quad G(3) = \frac{4\pi}{15}\ al\ (1 + 3/8\ \pi\ a/l) \qquad (3.3)$$

These factors calculated for the GM-counter 1 B 85 are plotted versus μ in Fig. 3.

Threefold coincidence telescope. The mounting of the GM-counter telescope and of the single counter is shown in Fig. 2. The counting rate of the telescope is given by

$$N_T = G_T(\mu)\ \varepsilon\ I_o \qquad (4.0)$$

If we denote by D the distance between the axis of the upper and lower counter and the diameter resp. the effective length of a single counter again by a and l we get for $G_T(\mu)$ after Greisen (L26) and Pomerantz (L27) :

Geometry Factors

$$\mu = 0: \quad \frac{n}{I_0} = a^2 \left\{ \frac{l}{D} \text{arctg} \frac{l}{D} + \frac{\pi a}{4D} \left[1 - \frac{D^2}{D^2 + l^2} \right] \right\} \left[1 + \frac{1}{2} \left(\frac{a}{D} \right)^2 \right] \quad (5.1)$$

$$\mu = 1: \quad \frac{n}{I_0} = \frac{2a^2}{3} \left\{ \frac{2(D^2 + l^2)^{1/2}}{D} - 1 - \frac{D}{[D^2 + l^2]^{1/2}} + \frac{\pi a}{4D} \left[1 - \left(\frac{D^2}{D^2 + l^2} \right)^{3/2} \right] \right\} \left[1 + \frac{1}{2} \left(\frac{a}{D} \right)^2 \right] \quad (5.2)$$

$$\mu = 2: \quad \frac{n}{I_0} = \frac{a^2}{4} \left\{ 3 \frac{l}{D} \text{arctg} \frac{l}{D} + \frac{l^2}{l^2 + D^2} + \frac{\pi a}{2D} \left[1 - \left(\frac{D^2}{D^2 + l^2} \right) \right] \right\} \left[1 + \frac{1}{2} \left(\frac{a}{D} \right)^2 \right] \quad (5.3)$$

$$\mu = 3: \quad \frac{n}{I_0} = \frac{2a^2}{5} \left\{ \frac{1}{D(l^2 + D^2)^{1/2}} (D^2 + 3l^2 - \frac{1}{3} \frac{l^4}{D^2 + l^2}) - 1 + \frac{\pi a}{4D} \left[1 - \left(\frac{D^2}{D^2 + l^2} \right)^{5/2} \right] \right\} \left[1 + \frac{1}{2} \left(\frac{a}{D} \right)^2 \right] \quad (5.4)$$

These factors for our telescope are also plotted versus μ in Fig. 3.

Efficiency. If τ is the sum of the dead time and recovery time of a counter and N_s its counting rate, we get for the probability that an ionizing particle passing the counter array causes a threefold coincidence

$$\varepsilon_T = e^{-3N_s \tau} \quad (6.0)$$

whereby we assumed $N_{s1} \approx N_{s2} \approx N_{s3}$ and $\tau_{s1} \approx \tau_{s2} \approx \tau_{s3}$ ($\tau \approx 100 \mu$ sec).
This can be considered as true in the limits of corrections which have to be applied in extreme cases.

Ionization chamber. The third detector of our balloon equipment is an integrating pulsed ionization chamber after Neher and Johnston (L25). It is essentially a steel walled sphere filled with argon of 9 atm. The thickness of the wall amounts to 0.6 mm corresponding to an absorbing layer of

480 mg/cm^2. The operation of the chamber is illustrated by Fig. 4. A central insulated collector-rod (a) is charged by the electrons and negative ions which are produced by the ionizing radiation. This causes an electric field to be build up between the rod and the inner cup (c) which is kept at the positive potential of the battery through the resistor R_1. Under the influence of the increasing field a gold coated quartz fiber (b), 10 μ thick and 0,5 cm in length, is attracted more and more to the central rod and finally contacts it after having tipped over from an instable position eventually reached. The rod is then discharged through the resistor R_2, the fiber flies off in the fieldless position to start for a new cycle. The voltage drop across R_2 during the discharging of the central rod provides a signal which is fed to an impedance converter and appropriate pulse shaping stages for the telemetering to ground (see Fig. 7 and 12). The peak of the signal amounts to 12 volts, the decay time to 10 μ sec.

For intercomparison of the flight data the different chambers were calibrated with a radium source in a standard geometry after the manufacturing and immediately before the flights. A third check is made during the flight. Since the ratio of the pulse rate of the ionization chamber and the counting rate of the single counter are of principal interest for the estimation of an average energy of X-Rays a normalization of this measure during each individual flight is most appropriate and can be utilized for calibration without referring to a special geometry. For this purpose we defined a normal ratio measured at an altitude where, according to experience, it is sufficiently constant except in the rare cases of very high energy solar proton events.

In practice we determined the time interval Δt_{11} between two discharge signals at the point where the counting rate of the single 1 B 85 counter just amounted to 11 c/s. Then the pulse rate I of the ionization chamber plotted in our presentation of results is defined by:

$$I = 0.62 \frac{\Delta t_{11}}{\Delta t} \text{ pulses/min}$$

where the time interval Δt refers to that measured at the different altitudes during the flight. The numerical constant 0.62 adjusts the unit to those used in previously published measurements. (L 16) The special standard counting rate of 11 c/s refers to an air pressure of 186 mb corresponding to an altitude of about 10 kms. The calibration point is not very critical. One can also make use of the point for 10 or 12 c/s or average over the region from 10 to 12 c/s and has then only to determine Δt_{11} by linear interpolation.

Temperature coefficient. The gold coated quartz fiber shows bimetallic properties though we took care to get an equally thick layer by turning the holding frame during the vaporizing of the gold. As to the orientation of the cylindrical surface of the fiber with respect to the plane in which it moves in the ionization chamber one gets a positive, zero, or negative temperature coefficient of the pulse rate at constant irradiation. It could be shown that by rotation of the same fiber around its axis the sign of the temperature coefficient could be changed in an interval of the rotation angle of about $2°$. This shows that in principle a practical zero coefficient can be achieved either by a little cumbersome adjustment or by selecting systems with accidental zero coefficients. We found it, however, more economical to use the systems without special adjustment or selection since the temperature of the chamber could easily be measured during the flights and hence also corrections for temperature change could be applied if necessary. In most cases, however, the variations of the temperature of the chamber could be limited enough to keep their influence negligeable. This was achieved by inserting the chamber in a protective envelope of styrofoam on which some stripes of black paper had been pasted to provide sufficient solar heating.

Energy threshold for detection of particles or photons. The energies of the charged particles and photons must exceed certain values in order to be recorded by the 3 different detectors. The threshold for charged particles is determined by the wall thickness of the detectors which must be smaller than the range of particles to be recorded. For photons such a threshold does not exist if single detectors like the GM-counter or the ionization chamber are involved.

An energy threshold exists, however, for the detection of photons by means of the counter telescope. This responds to secondary electrons released in the upper wall of the topmost counter with energies high enough to reach just the sensitive volume of the lowest counter. The various thresholds for the TESIO combination are compiled in table 2.

TABLE 2

Energy thresholds of the TESIO

	Single counter	Ionization chamber	Telescope
Electrons	160 keV	1.1 MeV	260 keV
Protons	3.8 MeV	16 MeV	9 MeV
Photons			260 keV

Relations between the counting rates of the single counter and the telescope bearing on the directional distribution of any radiation and the range and energy spectrum of charged particles. The geometry factors for the single counter and the telescope are plotted in Fig. 3 versus the exponent μ of the $\cos^{\mu}\vartheta$ law by which the zenith angle distribution of the measured radiation shall be approximated as usual.

In order to derive absolute flux densities from the counting rates of the single counter or the telescope μ must be known.

a. Range distribution

There is a possibility to obtain a rough estimate of μ if one has to deal with protons or charged particles of which the range spectrum $F(>R)$ can be approximated by a power law of the shape

$$F(>R) = C_1 R^{-m} \qquad (7.0)$$

R shall represent the range in g/cm^2 and C_1 and m certain constants. For particles which impinge at the top of the atmosphere and penetrate to an atmospheric depth corresponding to an air pressure P we can replace R in eq. (7.0) by P and express the vertical flux by

$$I_o(P) = C_1 P^{-m} \qquad (7.1)$$

With the assumption that the flux density (7.1) is isotropic at the top of the atmosphere we can write for that measured at P arriving from directions inclined towards the zenith by an angle ϑ replacing P by $P/\cos\vartheta$

$$I(P,\vartheta) = I_o\left(\frac{P}{\cos\vartheta}\right)^{-m} = I_o(P)\cos^m\vartheta \qquad (8.0)$$

hence
$$m = \mu$$

b. Ratio of counting rates

Another possibility to determine μ is based on the ratio of the counting rates of the single counter and the telescope. One promises that the efficiencies ε_s and ε_T for the radiation in question are inserted correctly in eq. (1.0) and (4.0), I_o calculated after both must be identical, hence it follows:

$$\frac{\varepsilon_T}{\varepsilon_s} \frac{N_s}{N_T} = \frac{G_s(\mu)}{G_T(\mu)} \qquad (9.0)$$

The ratio $\frac{G_s(\mu)}{G_T(\mu)}$ is plotted versus μ in Fig. 3. One recognizes that it depends strongly on μ. This can be utilized to determine μ by evaluating the term on the left hand side of (8.0) and find the corresponding abscisse μ in Fig. 3.

<u>Conversion of the range distribution of protons into an energy spectrum.</u> As already seen from eq. (7.0) and (7.1) the range distribution of protons can be obtained by measuring the flux as a function of the air pressure as it is done automatically during the ascent of the balloons. Since the energies of solar protons which can be measured in the auroral zone are usually in the order of 100 to 500 MeV their spectrum can easily be derived from the range distribution. We rewrite (7.1) in the form:

$$F(>P) = C_1 \cdot P^{-/\mu} \tag{7.3}$$

accentuating by taking the letter μ for the exponent its relation to the directional distribution after 8.0.
Then we insert for P

$$P = \left(\frac{E}{31}\right)^{1.8} \tag{10.0}$$

which is a relation for protons of kinetic energy E given by Brobek and Wilson (L28) and slightly modified for our purpose. It holds satisfactorily for

$$20 \text{ MeV} < E < 500 \text{ MeV}$$

We get from (7.3 and 10.0) the integral energy spectrum

$$F(>E) = C_1 \left(\frac{E}{31}\right)^{-1.8/\mu} \tag{7.4}$$

and by differentiating

$$\frac{dF(>E)}{dE} \equiv f(E) = -1.8/\mu \; 31^{1.8/\mu} \cdot C_1 \; E^{-(1+1.8/\mu)} \tag{7.5}$$

If we write the differential spectrum in the usual form:

$$f(E) \sim E^{-\gamma} \tag{7.6}$$

it is seen that a power law for the range distribution is converted into a differential energy spectrum simply by setting:

$$\gamma = 1 + 1.8\,\mu$$

<u>Relation between the counting rate of the single counter, the pulse rate of the ionization chamber and the average energy of X-Rays.</u> If it is known that only one kind of radiation causes a certain pulse rate of the ionization chamber and a certain counting rate of the single GM-counter the average energy of the corresponding quanta or particles can be estimated. Anderson (L18) has determined the ratio of the pulse rate of an ionization chamber practically identical with ours and of the counting rate of the 1 B 85 counters as a function of the energy of photons. The curve obtained by him was also utilized by us. It is drawn in Fig. 5, whereby the ratio is expressed in units of that, measured for the Galactic Cosmic Radiation at 10 mb air pressure. Though the curve is double-valued, in most cases of auroral X-Rays one can scarcely doubt that the lower value of the energy has to be taken. In other cases e.g. when photons in excess of the C. R. are still measured at air pressure levels as high as 50 mb, one has at first the choice whether an extremely high flux of photons with energies below 100 keV is present or a smaller flux of photons well above 100 keV which are not so strongly absorbed in the air like the former. In these relatively rare cases other criteria are needed to discriminate between these two possibilities. (see page 40)

V. <u>Circuitry of the TESIO Unit</u>

A block diagram of the electronic circuitry of the balloon equipment is shown in Fig. 7 and details of the various stages are given in Fig. 8 to Fig. 13. With exception of

the transmitter all branches are transistorized. The high
voltage supply is shown in Fig. 8, the subcarrier oscillators
in Fig. 13, the scaler stages in Fig. 10, the transmitter in
Fig. 13 and 15a, the pressure and temperature coding device
in Fig. 15b.

Some of the various stages and components have been described
in detail in our Technical Report No. 1 (L29). We can therefore restrict ourselves to point to the general lay-out of the
equipment and to stress only a few special features.

There are seven kinds of information to be transmitted:
The counts of the single counter, the coincidences of the
telescope, the pulse rate of the ionization chamber, the air
pressure as measured by a barograph covering the range of
1000 to about 100 mb with sufficient accuracy and another
one for the range between 200 mb to 4 mb, furthermore the
temperature of the counter and telescope component and the
temperature of the ionization chamber. Three of them, the
position of the pointers giving the two pressure information
and the temperature of the counter and telescope component
are transmitted by one subcarrier channel after having been
coded in a cyclic sequence of morse signs. This is achieved
by a rotating drum, driven by a motor, being a half cylinder
of aluminium plate with 600 grooves pressed in the surface
representing so 600 divisions in the direction of the axis.
(Fig. 15b)

If a pointer tip slides along a certain groove of the rotating
drum one subcarrier is switched on and off by a lac pattern
on the surface producing so morse signs corresponding to this
special one of the 600 divisions. Thus the positions of all
three pointers are transmitted once every half minute by this
device. It is commercially available as a compact unit.
(Fig. 15b) The information of the radiation detectors are
sequences of counts appropriately scaled down. The last stage

of the scaler switches the corresponding subcarrier on and off at a normal radiation level in approximately 20 seconds. The switching rate per unit time is proportional to the radiation intensity.

The temperature of the ionization chamber is measured by the temperature dependant resistance of a small semiconductor (Messleiter K17, 10 kOhms at 20° C, Siemens). This controls a phase shift generator Fig. 11 (not drawn in the block diagram of Fig. 7 and in the diagram of Fig. 13) changing the frequency in the range between about 4000 and 10.000 c/s if the temperature changes from +50° to -50° C.

The circuitry of the single counter and of the telescope is shown in the figures 8 and 9. The high voltage for the counters is supplied by a direct current converter consisting of a blocking oscillator and a cascade rectifier network. The oscillator-frequency is about 8 Kc/s. The voltage is stabilized by a Victoreen 5841 corona-discharge tube. It was necessary to test the voltage dependency on temperature of these tubes. Good ones showed a drop of less than 10 volts if the temperature changed from +20° C to -50° C. The total high voltage supply, the leak resistors, and the coupling condensors of the counters were enclosed in a screening box poured out with cold hardening Silicon Caoutchouc from Wacker-Chemie, München. The box with the counters is shown in Fig. 2. Other views of the box mounted together with the other components are given in Fig. 14.

To provide the high external leak resistance necessary for a sufficient life-time of the counters at extremely high radiation levels only a part of the output voltage of each counter is fed by the junctions X, Y, Z and W (Fig. 8 and 9) to individual amplifier stages. The coincidences are selected by a circuit of the Rossi-type with a directly connected discriminator. The actual resolving time is about 7 μ sec, quite enough for all cases if threefold coincidences are used.

The counting rate of the single counter is scaled down by 10, that of the telescope by 6 flip-flop-stages corresponding to scaling factors of 1024 and 64 respectively. Two successive flip-flop-stages are drawn in Fig. 10 showing the lay-out and the coupling of the stages. The output voltage changes between 0.6 and 5.8 volts and switches directly the sub-carrier oscillator (Fig. 13) the frequency of which can easily be adjusted by variation of only one resistor.

Four subcarrier units have been used with frequencies of 740 (pressure, temperature), 960 (single counter), 1300 (counter telescope), and 1700 c/s (ionization chamber). The fixed frequency oscillators and that one with varying frequency which gives the temperature of the ionization chamber, control after being mixed in a mixing stage, the capacity variation diode drawn in the lower left corner of Fig. 13. The latter one modulates the frequency of a 152 Mc/s Colpits-oscillator. Using a DC-70 valve a HF output of about 0.2 watt is obtained. This is necessary in favour of a long range if the balloons are floating far away from the receiving station. With the receiving equipment described below records were taken to a distance of about 500 kms. Fig. 14 and Fig. 15 show the transmitter. The $\lambda/2$-antenna is mounted at the lower side of the case in downward direction.

The ionization chamber channel (Fig. 12) is designed for a high input impedance (10 MΩ for 1000 c/s and 4 MΩ for 30.000 c/s at about 20° C). This is considered necessary in order to keep the load of the quartz fiber contact during the discharge as low as possible. The strong feed back is also of advantage for good performance at varying temperatures. If a signal from a source with 4 MΩ inner resistance is fed to the impedance converter the output decreases only 15 % between +30° C and -45° C if the electrolyte capacitors of the converter stage are of the tantalum type.

VI. Ground Station

The receiving equipment (Fig. 16) consists of a FM receiver (Nems Clarke Model 1502 A) which is fed by a ten element vertically polarized Yagi antenna adapted to 150 Mc/s with $\pm$ 5 Mc/s bandwith. The directional characteristic of the antenna has a half angle of about 20 degrees in the horizontal plane. Because of its directional properties the antenna cannot receive a transmitter which is overhead of the station. In this case a single folded dipol was alternatively used. The LF (low frequency) output is directly fed to an amplifier and after being mixed with time marking signals spaced by one minute to a tape recorder.

Following the former (Fig. 16) a combination of a high and low pass filter (cut off frequency each 4 Kc/s) separates the input in two channels. The higher one (frequency range 4 to 10 Kc/s) feeds a signal indicating the temperature of the ionization chamber by a corresponding frequency to the horizontal plates of an oscilloscope the vertical plates of which are connected with a normal RC Generator in order to produce Lissajou-Figures on the screen. A standing figure is reached by adjusting the frequency of the RC Generator appropriately which is then read out. This was done in intervals of 5 - 10 min. By this method a very accurate determination of the signal frequency is warranted even if the transmission is very noisy.

The low frequency channel ($<$ 4 Kc/s) is connected to four band pass filters of 740, 960, 1300 and 1700 c/s central frequency respectively and $\pm$ 7.5 % bandwith (3 db). The output of each filter is rectified and controls the grid of a cathode follower which feeds a corresponding channel of a multi-pen-recorder. In the rest state (no signal fed in) the saturation current flows; it is brought down to

practically zero by full signal amplitude. The response
between these two extreme states is logarithmic so that
also small signals cause recognizable traces.

VII. <u>Balloons and Accessories</u>

We employed polyethylene balloons manufactured by Raven
Industries with a volume of 80.000 cu ft. The balloons
were filled with hydrogen. The net weight of such a
balloon varies from about 17 to about 19 kg. The payload consisting of a TESIO, a parachute, lines, and a
squib for line cutting amounted to 8.1 kg at each flight.
These data determine a summit altitude reached by the
balloon of about 32 to 33 kms corresponding to pressure
levels of 7 to 8 mb respectively. In order that the
balloon ascends with a velocity of about 300 m/min which
proved to be a reasonable value, a free lift of about
20 % of the gross load has to be provided. This is obtained by filling the balloon on ground with about 36 m^3
of hydrogen. The accurate conditions were checked before
each flight and then the lift of the balloon weighed out
appropriately. Since the balloons were filled at open air
one has also to account for the dynamical lift when winds
are blowing. In the first place one has to keep the
balloon so stout that it remains always spherically. This
has the advantage that on the one hand the side wind
pressure does not increase to a troublesome extent even at
velocities up to 10 m/s and on the other hand the dynamical
lift can be estimated and compensated rather accurately for
this simple shape if the wind velocities are measured. The
filling procedure is performed in such a manner that a
clutch is mounted at a definite distance from the top of
the balloon confining so the spherical volume of the fully
inflated balloon. This clutch bears the weights. The
balloon is then filled through the inflation tube mounted
a little below the balloon's topside. For this purpose

the tube is put on a brass cylinder (15 cm diameter) to which
eight gas supply valves are soldered and connected to 8 hydrogen
bottles by means of rubber tubes. By opening always the next
bottle only after the preceding one had been emptied we could
fill the balloon in about 6 minutes. The fast inflation
causes the temperature of the gas to decrease by adiabatic
expansion to about $-30°$ C. If the balloon is then launched
without an appreciable delay the gas remains rather cold until
the ceiling altitude is reached where it warms up very slowly
by absorption of sunlight. The heat absorption of the gas just
compensates for the loss of gas by diffusion through the
polyethylene skin so that the floating level of the balloon
remains very constant. We had balloons which, after having been
filled in 6 minutes and launched two additional minutes later,
stayed at a level of 8 mb for more than 20 hours (during midnight sun time). Those, however, filled slowly in about 20 minutes and launched another 5 minutes later stayed first for
3 or 4 hours at the summit altitude but descended then slowly
to levels around 12 mb after 20 hours time of flight.

The launching of a balloon is performed by cutting the clutch
by which it was kept close to the ground before. The balloon
then rises freely drifting with the wind in such a manner that
the lines are just put under stress when it stands overhead
the payload. Then it begins to take off relatively smoothly
one piece of the payload after the other. The payload was
held up by a man in order to avoid bumping on ground.

Fig. 17 shows a drawing of the arrangement during the flight.
Below the balloon two boxes are mounted labeled by "squib".
This is the device for separating the payload from the balloon
if one of the following conditions holds:
1. The balloon stays during the ascent longer than one and
   a half hour below 12 kms (slow ascent due to a leak)
2. The balloon descends from the ceiling altitude and
   passes the 12 kms height niveau

3. The temperature inside the box goes below -20° C which is the limit where the safety squib still works reliably.

The squib is an iron cylinder in which a bolt is pushed through the bearing line by the explosion of a small charge of gun powder. The shot is ignited by electrical heating of a thin wire which is controlled by three switches (Fig. 18a):

1. a bimetallic thermo-switch
2. a clock switch connecting the battery with the control circuit 90 minutes after the launching
3. a baroswitch in series with the clock switch open below 200 mb and closed above.

This device is powered by two 2 volt dry (Rulag) accumulator cells. Photographs of the components are shown in Fig. 18b and Fig. 18c.

After firing of the squib the balloon bearing now only the "squib" boxes, rises so fast that it explodes in high altitudes while the TESIO descends on the parachute (silk, 2.4 m diameter). It hits the ground with a velocity of about 5 m/s. Photographs of the TESIO components ready for flight are shown in Fig. 18d.

VIII. Indications for Launching

Since we have been interested in measurements of unusual events we were depending on forecastings on account of various ground observations. Two kinds of warnings were provided by instruments of the Kiruna Geophysical Observatory.
1. By the riometer
2. By the magnetovariograph

Other very helpful warnings were kindly transmitted to us by NERA (Netherlands Postal and Telecommunications Services (courtesy of Dr. de Feiter), and by the Fernmeldetechnisches Zentralamt (courtesy of Mr. Ochs) on account of solar flare and ionospheric observations. The _riometer_ measures the cosmic noise in a narrow frequency band around 27 mc/s. If an additional ionization occurs between 60 and 100 kms altitude the cosmic noise in the above mentioned frequency band is most effectively absorbed and hence very sensitively indicated by the riometer. Additional ionization in these regions can be caused either by bremsstrahlung-X-Rays due to electrons braked in the region above this layer or by solar protons penetrating deeply into the atmosphere. The character of the records corresponding to the two sources of additional ionization is quite different so that it is often possible to discriminate clearly between both. In both cases, however, this instrument indicates only events of interest already in progress. This circumstance restricts the possibility to utilize it for forecasting purposes. For instance in the case of X-Ray activity showing mostly rapid variations sometimes only of short duration in the order of several minutes the riometer is not so appropriate because it may easily happen that the activity has ceased when the balloon reaches the ceiling.

A better warning instrument for X-Ray measurements is the _magnetovariograph_ if its record can immediately been read out. This possibility exists in Kiruna. Since bremsstrahlung-X-Rays are very often correlated with magnetic disturbances preceding the onset of the bursts by about 20 to 30 minutes it is advisable to make use of this instrument as indicator of events to be expected. Furthermore the magnetic activity shows very often recurrence tendencies which may also be utilized for the forecasting. Magnetic storms recur sometimes after 27 days. There is also a 24 hours' recurrence tendency observed very often in polar regions probably caused by a plasma cloud passing near the earth so that a local station can observe magnetic disturbance limited in time for several days (2 to 5) starting every day at about the same hour and showing every day a very similar amplitude pattern. This recurrence tendency is indeed a very good indicator for the expectation of X-Ray activity. This was utilized successfully in 1960 and 1961.

In order to catch solar proton events it is necessary to get information on flare activity as soon as possible. As already mentioned telegrams and phone calls from NERA and the Fernmeldetechnisches Zentralamt in Darmstadt were extremely helpful in this respect. The riometer also indicates strong flare activity by receiving solar radio noise. If such a noise storm is observed it is always advisable to launch a balloon.

In any case when the above mentioned indications are given the launching should be achieved with the shortest possible delay. The time needed by our group from the warning to the take off in most cases amounted to less than 30 minutes. In order to take advantage of every chance we watched the riometer and magnetometer all the time and kept always some instruments readily prepared for launching.

Immediately after the warning a short laboratory test of the equipment was effected by checking the counting rates of the detectors as well as the function of the electronic circuits. Thereafter the heat insulation was put on and the instruments, now ready for flight, were brought to the launching place in a distance from the observatory of about 500 m. Here a final test was carried out to check also the good performance of the transmission. Thereafter the inflation of the balloon started.

IX. Results

1. General Remarks

The results of the flights carried out in the period between June 12th and August 3rd with the TESIO units described in the preceding sections and with pilots (single counter units described in Technical Report No. 1) (L29) are compiled in Fig. 22 to 67. Most of the TESIOS - with exception of the units flown in SK 23, SK 30, and SK 31 - borne by the polyethylene balloons (Skyhooks abbr. SK in the flight schedule) floated at about 8 to 12 mb for 15 to 24 hours whereas the pilots, carried aloft with normal weather balloons, ascended in about 90 minutes only to a certain summit altitude around 30 kms where the balloons burst. Thereafter the equipments desceded to ground in about half an hour braked by appropriate parachutes. These latter flights served for probing either the radiation conditions at all or in case of the solar proton events for measuring the range distribution of the protons recorded at constant level by the TESIOS. All of the equipments were lost in the only weakly populated and relatively inaccessible regions of Lappland.

The schedule given by Fig. 20 shows the date and duration of our flights (SK and Pilots) and those of the flights at Minneapolis and Fort Churchill performed by the University of Minnesota Group. It is seen that especially in the period of July 12th to July 28th which was a very interesting one the flights in the three regions overlap satisfactorily.

In this report we must confine ourselves to present mainly the raw data of our measurements. A thorough analysis and discussion of the various events represents a work on its own which is in progress. The total contents of information can only be exhausted when all data referring to the associated phenomena of the events in question are available to us.

2. <u>Representation of Data</u>
The data are presented in a uniform manner.

a. <u>High Altitude Records</u>
The counting rates of the telescope, the single counter and the pulse rates of the ionization chamber are plotted on logarithmic scales with the same base in order to make their ratios for varying radiation conditions comparable. The fixed scale is adjusted to give immediately the counting rate of the single counter. For fitting it to the coincidence rate of the telescope and the pulse rate of the ionization chamber special marks are inserted in the figure.

In order to make visible at a glance what kind of events manifest themselves in the 3 curves, the following should be kept in mind:

The interesting radiation fluxes are superimposed to the normal Galactic C. R. and must be evaluated by subtracting this background. If there are only bursts of excess radiation when the balloon floats at ceiling the latter is recognized without ambiguity. In case of long lasting events it must be estimated by extrapolation from the readings at ground station monitors as shown by Ehmert, Erbe and Pfotzer (L30).

In order to indicate the expectation levels of the Galactic C. R. the plateau values are marked by dashes for cases where excess radiation was recorded. The dashes are labeled by CN, IN and TN for the single counter, the ionization chamber and the telescope respectively.

During the ascent one has also to take into account the shape of the cosmic ray intensity versus pressure curve. which is different if measured by the telescope, the single counter, and the ionization chamber. This can be understood as follows:
A beam of Galactic C. R. particles pehetrating into the atmosphere produces secondaries and is at the same time attenuated due to this process. This causes the total

particle flux to increase until a layer equivalent to
~ 90 g/cm^2 (air pressure 90 mb) is traversed. At this depth
the absorption of primaries equals the production of secondaries. Then with increasing depth the absorption exceeds
the production and the particle flux decreases. Vice versa
the counting rates of the telescope and the single counter
increase first during the ascent until the secondary maximum
is reached, then decrease to a finite value for the primaries
at zero air pressure. Since the single counter detects
particles from all directions in the upper hemisphere the
maximum is smeared out and less pronounced than for the
telescope which responds essentially to particles from
directions close to the vertical. In addition it appears
at smaller pressure (~ 40 - 50 mb).

In contrast to the detectors which count single particles
the ionization chamber measures the product of the particle
flux and the average ionization of the particles in the
chamber. Since towards high altitudes the percentage of
slow particles which ionize stronger increases, the decrease
of the total flux is just balanced by the increase of
specific ionization. Therefore the pulse rate of the
ionization chamber in northern latitudes does not pass a
maximum but gradually approaches a plateau at 40 mb.

The excess radiation can be classified to consist of photons
when single counter and ionization chamber do respond and
the counting rate of the telescope is not affected. Charged
particles of lower energies than that of the Galactic C. R.
are indicated if the relative response of the telescope
and the ionization chamber is higher than that of the single
counter. This is due to the stronger absorption of the low
energy particles arriving from inclined directions and on
their stronger specific ionization. The first effect causes
a relatively higher loss of particles for the single counter
with respect to the telescope, the latter an over-balancing

of this loss for the ionization chamber.

b. <u>Riometer</u>

The curves labeled by riometer represent the intensity of cosmic noise at 27 Mc/s as measured by the riometer at Kiruna.[+] They correspond to the direct pen records of the instrument redrawn on logarithmic scale. No correction for non-linearities of the instrument was applied because this approximation is quite sufficient for our purpose. The broken line corresponds to the normal level of cosmic noise and hence the difference between the two curves is a measure of the cosmic noise absorption (CNA) approximately in decibel. Since the irregular absorption is caused by the ionization of an atmospheric layer mainly between 50 and 100 kms height by ionizing radiations in excess of the Galactic C. R., it is an indirect measure of this radiation flux which in many cases can also be detected directly by balloon equipments in altitudes down to 30 kms. It is for this correspondence that we plotted the CNA just opposite as usual in order to demonstrate also clearly in the diagram that increasing CNA means increasing intensity of ionizing radiation.

c. <u>Magnetic Components</u>[+]

The curves labeled by ΔH_x, ΔH_y, and ΔH_z represent the deviation of the magnetic field components on ground from the normal values H_x, H_y, and H_z at Kiruna marked by the broken lines.

H_x is the horizontal component pointing to the north, H_y that pointing to the east and H_z the vertical component pointing downwards. In order to get an idea on the direction

---

[+] These data were kindly put at our disposal for the purpose of this report by Dr. B. Hultqvist and Dr. J. Ortner, Kiruna Geophysical Observatory. If there should have been introduced any errors caused by outworking of the original records only the authors are responsible. See, however, also the concluding remarks on page 45.

of the current system flowing in the ionosphere which causes the magnetic disturbance vector one has only to turn the magnetic components in such a manner that ΔH_x points to the east and ΔH_y to the south, then the horizontal resultant points in the direction of the ionospheric current.

Since normally $\Delta H_x > \Delta H_y$ the following rules hold if we assume a ribbon like current in the ionosphere.

$$+ \Delta H_x \brace + \Delta H_z$$ current west - east, north of station

$$- \Delta H_x \brace + \Delta H_z$$ current east - west, south of station

$$- \Delta H_x \brace - \Delta H_z$$ current east - west, north of station

$$+ \Delta H_x \brace - \Delta H_z$$ current west - east, south of station

$\Delta H_z = 0$ current overhead

3. **Comments on the Various Flights**

In the following some indications to the remarkable features of the various flight records shall be given. For this purpose we subdivide the records into two groups:

a) The first one embraces those events which are clearly not related to the solar flare effects between July 12th and July 28th

b) The second one refers to those just occurring in this very period. (Entirely normal flights with only Galactic C. R. present will be omitted).

Group a) **Flights not correlated with outstanding solar flare effects**

Flight SK 12 (Fig. 22, 23, 24) This was performed on a

routine base without special warning. According to that the records of all instruments at first are completely normal from 0230 UT on June 11th until 0230 UT on June 12th. Thereupon a gradual increase of Cosmic noise absorption (CNA) begins, followed by a steplike one from 1 to 3 db between about 0530 and 0540 UT (the gap, not hatched, indicates that no record is available). The step is reflected by the influx of X-Rays in balloon altitudes present for 1.5 hours. In contrast to this no indication of correlated magnetic disturbances is recognized. A second weak response of the riometer begins on June 12th at 11 UT lasting for about six hours which has now also a weak counterpart in ΔH_x but not in the counting rates of the balloon-borne instruments.

Flight SK 13 (Fig. 25, 26) This was launched on June 15th at 1716 UT because the CNA at 16 UT increased recognizably from about 0.3 to 0.7 db. This was considered to be the possible beginning of more pronounced disturbances following a rather quiet period. The forecast of detectable X-Rays based hereupon proved indeed to be true in that two bursts occurred between 15 d at 24 UT and 16 d [+] at 0110 UT. They are nicely correlated to the riometer record and also to the shape of a magnetic bay. Unfortunately the power of the mains at the ground station went off after the second peak so that nothing can be said about the further development of this event.

Flight SK 15 (Fig. 27, 28, 29) In the night of June 20th to 21st a pronounced CNA, peaked at about 20 d at 22 UT and 21 d at 02 to 03 UT (3.2 db), was observed and considerable magnetic disturbances developed from about 20 d at 19 UT to a deviation from normal between 400 and 660 γ in the period of 21 d from 00 to 03 UT. Since by experience a recurrence of similar

---

[+] For brevity the letter d is affixed here and in the following where it is convenient to indicate that the preceding date denotes the day of the month in question.

events was to be expected 24 hours later a balloon was launched in the following evening at 1904 UT. This proved again to be true. Around midnight on 21/22 d strong magnetic bays began to develop and outstanding bursts of X-Rays were observed on 22 d between 00 and 01 UT, and between 03 and 05 UT, and smaller ones around 02 UT, between 07 to 09 UT and 11 and 12 UT. This is to be considered again as an example of the 24 hours' recurrence type as previously described by the authors.

Flight SK 16 (Fig. 30) This flight was launched about one hour after the transmission of flight SK 15 on July 22nd had faded out because the magnetic disturbances were still on an unusual level and a slight increase of the CNA was in progress. We thought that perhaps a third recurrence of the influence which caused the event in the night of June 20/21 would occur. This seems indeed to be indicated by the negative bay centered around 23 d at 01 UT, the increase of CNA and a slight indication of X-Rays reaching an atmospheric level at 10 mb as also recorded by the GM-counter (the ionization chamber malfunctioned during a part of this flight). This third event of a presumably interrelated series was, however, rather weak.

Flight SK 19 (Fig. 31, 32) In the night of July 4/5 a considerable magnetic storm (deviations in the horizontal components between 400 and 660 γ) accompanied by a CNA was observed and hence X-Rays were to be expected after about 24 hrs. These occurred indeed. There are clear magnetic bays correlated with the X-Ray flux between 5 d at 21 UT and 6 d at 0130 UT but in contrast to this practically no magnetic correspondence was found to the X-Ray flux measured between 6 d at 0230 UT and 6 d at 0430 UT.

Flight SK 20 (Fig. 33, 34) On July 9th at 09 UT a gradual increase of the CNA was observed which was not accompanied by magnetic effects. Since this is characteristic for proton injections a flight was launched at 1051 UT. Though a weak increase of the CNA is indicated between 11 and 14 UT no correspondence can be found in the records of the three high altitude detectors.

Flight SK 34 (Fig. 65, 66, 67) The horizontal field components changed strongly in the quarter hourly interval centered around 0015 UT on August 2nd and deviations of nearly 1000 γ were observed from 0030 until 0130 UT. In addition the riometer showed the presence of auroral CNA. This gave rise to launch flight 34. When the balloon reached ceiling at about 0345 UT the disturbances had decreased to an almost insignificant amount. At 0415 UT, however, all instruments indicated the passing of a new disturbance. The onset of a negative magnetic bay in H_x at about 04 UT is clearly correlated with the appearance of a strongly fluctuating X-Ray flux also indicated by an increase of CNA from 1 to about 4 db at 0430 UT. The X-Ray flux continued until 0930 UT whereas the magnetic bay lasted only until about 07 UT. A positive magnetic bay observed between 12 and 19 UT had no counterpart in the high altitude and ground level record of the detectors sensitive to ionizing radiation.

Group b) **Flights during the period of enhanced solar activity in July 1961**

Between July 12th and July 28th there was a period of enhanced solar activity similar in character to that of July 1959. The activity was mainly due to the two McMath plage regions 6171 and 6178 and in one case (15 d) also the region 6172 could have been involved in the solar terrestrial phenomena we are interested in. The McMath plage region 6171 appeared at the eastern limb of the sun on July 8 to 9 with a sunspot group in the early stage of its development (C_9 according to the map of the sun of the Fraunhofer Institut, Freiburg [+]).

[+] Letter = state of development after the Zürich classification (Publ. der Eidgenössischen Sternwarte Zürich, Bd. IX, Heft 1) digit = number of spots

It passed the southern hemisphere at about 7 degrees in the average and disappeared behind the western limb on July 20th. The sunspot group reached its maximum development with 38 single spots when passing the central meridian on July 13th in the state E. Plage region 6172 with a spot group in the H state passed the disk between July 10th and July 23rd exhibiting a flare 3^+ at 15 E at about $15°$ N. We shall see below that it cannot yet be decided whether this flare or another one only of importance 2 in the very active region 6171 or possibly both have to be considered as responsible for the geomagnetic storm on 18 d. The plage region 7178 was seen first at the eastern limb on July 19th. It contained a class H sunspot group during its passage of the disk developing from 4 single spots to 22 spots on July 27th at about $30°$N. The western limb was reached on July 31st when the number of spots had diminished to 3 in the end state I of the group. The region was located between about 8 to $13°$ N.

During the passage of these three plage regions four cases of solar proton injection occurred (12 d, 18 d, 20 d, 28 d). Their developments were recorded by means of nine flights with TESIOS of 15 to 20 hours duration and these were supplemented by five additional flights of short duration in part for determining the absorption curve of the excess radiation (Fig. 56, 57, 58).

Table 3 shows a compilation of the solar and terrestrial phenomena connected by lines as they were very probably linked to each other. This table is supplemented by Fig. 21 showing the correlations between magnetic storms, solar flares, Galactic and Cosmic Ray data. It shows furthermore the duration of the various flights performed by the Minneapolis and our group and the kind of events measured by our TESIOS.

The following comments on the various phenomena are on the general lines of a paper presented at the Cospar Meeting 1962 in Washington by the authors. (L31)

T A B L E 3

Interrelated events in the period between July 11 and July 28, 1962

Date	Time UT	Event	Position	McMath plage region	Remarks
11.7.1961	1615 – 2040	Flare 3$^+$	06S 32E	6171	
	1650 – 1750	→ SCNA 3			no protons
	1651	→ SID			
	1655 – 1845	→ Type IV			Davis
	1702 – 2300	→ Type IV			HAO Boulder
12.7.1961	1000 – 1330	Flare 3$^+$	07S 23E	6171	protons observed at balloon altitudes injection already going on at 1945 UT outstanding
	1003	→ SID (5.5 hrs)			
	1020	→ Sfe			
		→ Type IV			
	1020 – 1133	→ SCNA			
13.7.1961	1113	→ SSC magnetic storm max K_p = 8$^+$			SSC-X-Ray event
	1312	→ Forbush decrease			proton intensity decreasing X-Ray events of various character persisting during the whole recovery time of Galactic radiation until July 18th
14.7.1961	0812	SI magnetic storm max K_p = 8$^+$			
	08 – 10	→ Forbush decrease			

Date	Time	Event	Coordinates	Region	Location/Notes
15.7.1961	1433 – 1929	Flare 3	14N 14E	6172	
	1508 – 1549	Flare 2	07S 21W	6171	
	1522 – 1803	Type IV			HAO Boulder
17.7.1961	1826	SSC			
18.7.1961	0000	Forbush decrease onset			
	0921 – 1330	Flare 3^+	06S 56W	6171	
	0940	SID			
	1000	Injection of rel. high energy protons begins			f.i. Deep River and Sulphur Mountain Neutrons
	1030	Injection of rel. low energy protons begins			Kiruna, balloon altitudes
19.7.1961	0940	Solar Protons still present			Kiruna, balloon altitudes
20.7.1961	0248	SSC magnetic storm max $K_p = 6^-$			
	1525 – 1726	Flare 3	05S 90W	6171	
	1552 – 1730	Type IV			HAO Boulder Davis
	1600	Injection of solar protons			f.il Deep River Neutrons
	1745 – 1748	Type IV			HAO Boulder
	1828 – 1942	Flare 3^+	07S 90W	6171	

Date	Time	Event	Location	Region	Notes
21.7.1961	0000	Rel. low-energy solar protons present			Kiruna, balloon altitudes
	1736 – 1755	Flare 3	01S 90W	6171	
	1800	No detectable proton intensity			Kiruna, balloon altitudes
	1936 – 0148	Type IV			HAO Boulder 4 Bursts
24.7.1961	0403 – 0640	Flare 3	13N 18E	6178	
	1722 – 2214	Flare 3	07N 10E	6178	
26.7.1961	1950	SSC magnetic storm max $K_p = 8^+$			
28.7.1961	0000	Low energy X-Rays (70 keV)			Kiruna, balloon altitudes
	0240 – ?	Flare 2$^+$	13N 40W	6178	
	0300	Injection of rel. low energy protons begins			Kiruna, balloon altitudes superimposed are X-Ray bursts of variable character
	1512 – 1948	Flare 3	08N 44W	6178	

Some viewpoints refer to papers presented by Hoffman and Winckler (L32) on their results during a part of the simultaneous flights at Fort Churchill and Minneapolis and also on a paper of Pieper, Zmuda and Bostrom (L33) concerning measurements with the University of Ionwa Injun satellite (1961 Omicron 2) on July 13th.

Events around July 11th, 12th and 13th

<u>A first sequence of events</u> interfering with each other started with a solar flare of importance 3 on July 11th at 1615 - 2040 UT. This was accompanied by an SID and Type IV radio emission. About 3 hours after the start of the flare a smooth increase in counting rate at high altitude was observed (Fig. 36) which reached at 22 UT a value corresponding to a flux of 0.06 protons/cm^{2}sec ster (at balloon altitude). The proton energy exceeded the 270 MeV range value (15 g/cm^2); the relative response of the different detectors points also to a high energy event. It seems that no remarkable magnetic solar flare effect (Sfe) has been recorded (Bartels private comm.). This flare was significant with respect to the following events in that the plasma beam originating from it caused very probably the SSC magnetic storm and the Forbush decrease on July 13th. A second flare of importance 3^+ occurred at 1000 UT on July 12th and was observed until 1330 UT. The Sfe and SID (Sudden ionospheric disturbance) associated with it were outstanding but the Type IV radio emission was - as far as we know - only relatively weak (1^-) (CRPL-Boulder, part B, F 204, 1961). The flare caused proton injection which was recorded first by our balloon-borne detectors when passing a level of 35 mb (Fig. 37). A 10 mb ceiling level was reached at 1945 UT. At that time the injection was probably already in progress for several hours. Assuming protons to be the only kind of excess radiation the range distribution indicates an energy spectrum of the shape

$$f(E)dE \sim E^{-4.4 \pm 0.35}dE \qquad 100 \leq E \leq 220 \text{ MeV}$$

The counting rates of the telescope, the ionization chamber, and the single counter were respectively 2.6, 2.7, and 1.6 times the normal value at 10 mb. These data correspond to an absolute flux of about <u>0.6 solar prot./cm^2 sec ster</u> in the above mentioned energy limits. The flux remained rather steady until July 13th at 03 UT (Fig. 38) and then decreased gradually to 0.2 prot./cm^2 sec ster towards the end of the flight at about 1120 UT on July 13th.

At 1113 UT the above mentioned magnetic storm sudden commencement and at 1312 UT the onset of a deep Forbush decrease of the Galactic Cosmic Radiation were recorded as a consequence of the July 11th flare. The arrival of the plasma beam at the earth was accompanied by a strong radiation burst coinciding with the SSC and detected by our balloon-borne apparatuses as well as by the Kiruna riometer. Events of this special type have been observed until now only twice (L18, L35). They were interpreted as to be due to X-Rays with energies in the order of 40 - 60 keV, however, in the light of the following considerations one can also think of χ-Rays with energies of about 700 keV. A thorough discussion of this special event will be published elsewhere (L34). Coinciding with the onset of the Forbush decrease at 1312 UT two small increases in radiation were measured with a single counter over Lindau/Germany (compare table 3). The two peaks shown in Fig. 39 seem to be quite outside of statistical errors.

Approximately four hours later (13 d at 1530 UT) when a new balloon had been launched (Fig. 40, 41) probably nuclear Gamma-Rays were detected at atmospheric depths $\leq$ 50 mb as well as a diminishing flux of excessive protons of about 0.1 - 0.2 per cm^2 sec ster below 7 mb ($E_{kin} \approx$ 90 MeV). Since this flux is of the same magnitude as prior to the onset of the Forbush decrease it can be inferred that the plasma cloud enclosing the earth at that time <u>contained only few trapped protons</u> in this energy range.

The bump in the record of the telescope between 16 and 17 UT could have been due to Compton- or photo-electrons released in the upper counter of the telescope by high energy Gamma-Rays of origin discussed below.

The flux of photons indicated by the ionization chamber and the single counter showed very peculiar and rapid fluctuations resembling in part a switching on and off correlated with magnetic disturbances. This may reflect time variations of a primary proton component of relatively low energy which could not penetrate down to a pressure level of 10 mb but manifest itself by releasing of nuclear Gamma-Rays.

This conclusion is based on a comparison of our flight data with those of the Fort Churchill flight M 307S as kindly communicated to us by Prof. Winckler during the COSPAR Meeting 1962. The following reasoning may substantiate it:

About 4 hours after the SSC burst the records of our single counter and ionization chamber show structures (Fig. 40) which appear simultaneously also in the measurements over Fort Churchill. Comparing both, it seems that the pattern which occurred between 1704 and 1910 UT (Fig. 40) was a repetition of a completely similar one occuring before 1615 UT and measured over Fort Churchill. Probably the end of this structure is also seen in our record whereas during the preceding part our balloon, still ascending, was not yet high enough. Assuming this to be true, we measured then during the ascent Gamma-Rays which penetrated to about 50 mb.

Since on account of the double valued calibration curve we have the choice to attribute energies to the measured photons either at about 70 keV or 700 keV we must discriminate against this ambiguity because the event exhibited rather unusual features in comparison to the more common ones in group (a). If we assume now that we measured the same kind of radiation as detected by the scintillation counter of the University of

Minnesota Group over Fort Churchill we can also adapt their findings that the photons must have been in an energy range higher than 100 keV. Hence the value of 700 keV should be preferred. Our assumption that the Gamma-Rays may be due to nuclear interaction of protons is supported by measurements of the State University of Iowa's Injun satellite (1961 Omicron 2) published by Pieper, Zmuda and Bostrom (L33). This satellite detected fluxes of protons with energies between 1 and 15 MeV as high as $3,3 \cdot 10^4$ $cm^{-2}sec^{-1}sterad^{-1}$ on July 13th. The energy spectrum was very steep as can be inferred from the fact that at the same time the flux of protons with energies $\geq$ 40 MeV was in the order of 10 to 20 $cm^{-2}sec^{-1}sterad^{-1}$.

We believe therefore that the bulk of these protons could not penetrate to balloon altitudes but produced Gamma-Rays probably by Coulomb excitation which then could be detected by the balloon apparatuses in Kiruna and Fort Churchill. Unfortunately the data about cross sections for radiative interactions in this region are too meagre at present for ascertaining that our assumptions are correct. A paper of Wakatsuki et al. (L36) concerning cross sections for radiative interactions of 14 MeV protons at first glance does not seem to contradict completely our interpretation. If this conception of nuclear Gamma-Rays should prove to be correct on account of other correlated data or by future measurements it would imply that probably also the Sudden Commencement effect would have to be interpreted on the same lines rather than in terms of auroral X-Rays. Unfortunately these questions cannot yet be answered without ambiguity.

Concurring with the peculiar fluctuations of the photon component the flux of more energetic solar protons faded out gradually and was essentially zero after 13 d at 22 UT. The gamma flux persisted, however, until 23 UT then its character changed showing a time pattern similar to the riometer record as usually seen, surviving the disappearance of the magnetic disturbances at 03 UT by approximately two hours.

The following magnetically nearly quiet period is terminated by a Sudden Impuls (SI) at the beginning of the second violent magnetic storm at 0812 UT on July 14th (Fig. 21). We believe that the latter was induced by the plasma beam of the Cosmic Ray flare on July 12th. This caused again a depression of the Galactic Cosmic Ray flux and at the same time a nearly continuous flux probably of X-Rays of lower energies (below 100 keV) appeared at balloon altitudes (intensity: 28 photons/cm^2sec ster) (Fig. 42). The latter persisted for about 5 hours without exhibiting a correlation with the pronounced variations of the magnetic field components, but a close one with the riometer record is shown; it was followed in a later phase by two short time bursts well correlated with magnetic field and riometer variations as usually seen in the auroral zone (Fig. 43). (The peak intensities at about 1830 and 2015 UT on July 14th amounted to 200 photons/cm^2sec sterad and 130 photons/cm^2sec sterad respectively).

Similar conditions prevailed during the recovery phase of Galactic Cosmic Radiation until July 18th (Fig. 44, 45, 46, 47, 48). Remarkably grouped spikes of X-Rays were observed on July 16th between 0030 and 0300 UT (Fig. 44, 45). They have counterparts in very high spikes observed at Fort Churchill by Bhavsar (University Minnesota Group, Kyoto Conf. 1961) and at Minneapolis appearing just in a rather quiet period between two groups of bursts at Kiruna. The X-Ray peaks at Kiruna were in the order of 1.5 to $1.8 \cdot 10^3$ photons/cm^2sec whereas the University of Minnesota Group measured a peak of $1.8 \cdot 10^5$ photons/cm^2sec of an average energy of 50 keV.

The records of these events can be considered as a further contribution to the morphology of X-Ray bursts in the auroral zone and shall be analysed together with those of group (a) in a later publication in all details.

The events of July 18th and 20th

A second interelated series of events started with the flares on July 15th. There are two flares, one of importance 3^+, in McMath plage region 6172, the other of importance 2 in plage region 6171. As already mentioned we cannot yet decide which of both caused the magnetic storm on July 17th [+)] that started with a SSC at 1826 UT and was followed by a Forbush decrease of the Cosmic Radiation in the early morning of July 18th. Shortly after the onset of the decrease solar protons were injected originating from a flare 3^+ occurring between 0921 - 1330 UT on July 18th. It is noteworthy to mention that if the beam which caused the Forbush decrease stemmed from the 6172 flare this should be an essential point in judging the propagation conditions for the solar proton injection on July 18th.

The increase of counting rates (Fig. 49) at an atmospheric level of 40 mb over Kiruna can be fixed to 1030 UT that of the neutron counting rate at Deep River at 1000 UT (Carmichael and Steljes, 1961). Whether this is a dispersion or an impact zone effect is not yet clear.

The radiation flux as measured by the telescope and the single counter shows at the beginning a quasiperiodic structure. The following characteristics of the protons could be inferred:

1. The range distribution becomes steeper with increasing atmospheric depths. If the counting rate n(p) is represented by a potential function of the pressure p

$$n(>p) \sim p^{-/u}$$

   one gets for
   - 240 mb $\geq p \geq$ 120 mb $/u \approx 3.7$
   - 120 mb $\geq p \geq$ 60 mb $/u \approx 2.9$
   - 60 mb $\geq p \geq$ 20 mb $/u \approx 2.1$

   This corresponds to an energy spectrum of

   $$f(E)\, dE \sim E^{-\gamma}\, dE$$

[+)] New aspects concerning the coherence of both arose from discussions during the COSPAR Meeting in Washington, 1962, especially by remarks of H. Carmichael. These will be discussed elsewhere.

for
$$600 \text{ MeV} \geq E \geq 420 \text{ MeV} \quad \gamma \approx 7.7$$
$$420 \text{ MeV} \geq E \geq 290 \text{ MeV} \quad \gamma \approx 6.2$$
$$290 \text{ MeV} \geq E \geq 170 \text{ MeV} \quad \gamma \approx 4.8$$

2. Maximum proton flux measured at 1530 UT at 10 mb
   ($E > 125$ MeV) : 30/cm^{2}sec sterad

3. The flux was decreasing $\sim t^{-2}$, if $t = \tau - \tau_o$
   τ = time, τ_o = time of the onset of the flare.

The flare 3$^+$ on July 18th caused probably the SSC at 0248 UT on July 20th (Fig. 21). During the moderate magnetic storm a flare 3$^+$ (1525 - 1726 UT) in the same plage region injected solar protons as for instance observed at Deep River and Sulphur Mountain (Boulder Report CRPL Part B) between 16 and 18 UT.

At midnight on July 20/21 low energy protons were still measured at balloon altitudes in Kiruna (Fig. 58) but it cannot be excluded that they stem from a second flare 3$^+$ at 1828 - 1942 UT on July 20th (Fig. 21). The energy spectrum can be approximated by
$$f(E) \, dE \sim E^{-5.5 \pm 0.5} \, dE \quad ; \quad 400 \text{ MeV} \geq E \geq 100 \text{ MeV}$$

The event on July 28th

The SSC storm extending from July 26 at 1950 UT to the early morning on July 28 is attributed by us to the flare 3$^+$ on July 24 at 1722 - 2214 UT in plage region McMath 6178 near the center of the solar disk. It was associated with a large Forbush decrease following the SSC. During the latter a balloon was launched at Kiruna (Fig. 62, 63). At a ceiling of 10 mb at first only X-Ray activity was recorded around mignight on July 27/28 (Fig. 62). At 0300 UT on July 28, however, proton injection was recorded by all instruments but also X-Rays were superimposed to the proton flux (Fig. 64).

This injection is tentatively attributed to a flare 2$^+$ also in McMath plage region 6178 firstly observed at 0240 UT on July 28 (after Mitaka and Kyoto observations).

Taking this time as τ_o and plotting the coincidence rate $N(t)$ of the telescope versus $t = \tau - \tau_o$ we found clearly
$$N(t) \sim t^{-0.8 \pm 0.1}$$

This report is a preliminary one and refers only to the rough data derived without ambiguity mainly from our own records and from the routine publications of geomagnetic data. A thorough and detailed analysis together with a discussion of the ground observations of the geomagnetic field variations and the Cosmic noise absorption (Riometer) will be published later in a joint paper together with our Swedish colleagues. Since most of the Kiruna flights were performed simultaneously with flights of the University of Minnesota Group at Fort Churchill and Minneapolis further valuable information will be available after all measurements will be analysed and compared.

Acknowledgement

We are deeply indebted to the Royal Swedish Academy of Science for permission to carry out balloon launchings at Kiruna Geophysical Observatory. We express our gratitude to the director of the Observatory Dr. B. Hultqvist and all colleagues and members of the Observatory staff for the cordial and friendly cooperation.

To Prof. J. R. Winckler, University of Minnesota, School of Physics, Minneapolis, and his group and to Prof. R. R. Brown, University of California, Berkeley, we are also very indebted for the successful interchange of information and warnings during the summer period 1961 and for all the helpful discussions of the mutual preliminary results.

Furthermore we might express our gratitude to Dr. L. D. de Feiter, Netherlands Postal and Telecommunications Services, Dr. Neher Laboratory, Leidschendam, and to the Fernmeldetechnisches Zentralamt der Deutschen Bundespost in Darmstadt, especially to Dipl.-Phys. A. Ochs, for supplying us with helpful information on the solar activity.

Finally we appreciate the well concentrated efforts of all the technical staff of our institute in preparing and carrying out the measurements and in processing the numerous data.

The research reported in this document has been sponsored in part by the OFFICE OF SCIENTIFIC RESEARCH, OAR, through the European Office, of the Office of Aerospace Research, United States Air Force and by the MAX-PLANCK-GESELLSCHAFT.

Literature

L1 Ginzburg, V.L.
Progress in Elementary Particle and Cosmic Ray Physics IV
ed. by J.G.Wilson and S.A. Wouthuysen, North-Holland Publishing Comp., Amsterdam 1958,
p. 348 - 359

L2 Sekido, Y., I. Kondo, T. Murayama, Y. Kamiya, S. Sagisaka, H. Keno, S. Mori, H. Okuda, T. Makino, and S. Sakakibara
Observations of the Point Source of Cosmic Rays
J. of Phys. Soc. of Japan $\underline{17}$, Suppl. A III, 1962, Part III,
p. 131 - 139

L3 McCracken, K.G.
The Cosmic Ray Flare Effect
J. Geophys. Res. $\underline{67}$, 423-458, 1962

L4 Obayashi, T. and Y. Hakura
Propagation of Solar Cosmic Rays through the Interplanetary Magnetic Field
Report of Ionosphere and Space Research in Japan, Vol XIV, NO. 4
1960

L5 Pfotzer, G.
Protonenstürme im Interplanetaren Raum, Review Article
Die Umschau in Wissenschaft und Technik, Heft 6, p. 178-181, und Heft 7, p. 197-201, 1962

L6 Ehmert, A. a. G. Pfotzer   Registrierungen der Kosmischen
                                Strahlung in der Zeit vom 12. -
                                15. November 1960,
                                Beitrag in den Mitt. a.d. Max-
                                Planck-Institut für Aeronomie,
                                Nr. 8, 1962

L7 Tobias, C.A. a. R. Wallace
                                Particulate Radiation: Electron
                                and Protons up to Carbon
                                in Medical and Biological Aspects
                                of the Energies of Space
                                ed. by P. A. Campbell
                                Columbia University Press, New
                                York and London 1961, p. 421-442

L8 Van Allen, J.A.   Communication to
                                Science News Letter 30, 287,
                                October 28, 1961

L9 Masley, A.J. and A. D. Goedeke
                                Complete Dose Analysis of the
                                November 12, 1960 Solar Cosmic
                                Ray Event
                                3rd COSPAR International Space
                                Science Symposium, Washington D.C.
                                April 30 to May 9, 1962
                                (to be published in Space Res. III
                                North Holland Publishing Comp.)

L10 Meredith, L.H., M.B. Gottlieb and J.A. Van Allen
                                Direct Detection of Soft Radia-
                                tion above 50 kms in the Auroral
                                Zone
                                Phys. Rev. 97, 201-205, 1955

L11 Van Allen, J.A.   Interpretation of Soft Radiation
                                    Observed at High Altitudes in
                                    Northern Latitudes (Abstract)
                                    Phys. Rev. 99, 609, 1955

L12 Van Allen, J.A.   Direct Detection of Auroral
                                    Radiation with Rocket Equipment
                                    Proc. Natl. Acad. Sci. US, 43,
                                    57 - 62, 1957

L13 Davis, L.R., O.E. Berg, and L.H. Meredith
                                    Direct Measurements of Particle
                                    Fluxes in and near Auroras,
                                    Space Research, ed. by H. Kallmann
                                    pp. 721-735, North Holland
                                    Publishing Company 1960

L14 McIlwain, C.E.   Direct Measurements of Protons
                                    and Electrons in Visible Aurorae.
                                    Space Research, ed. by H. Kallmann
                                    pp. 715-720, North Holland
                                    Publishing Company 1960

L15 Van Allen, J.A., G.H. Ludwig, E.C. Ray and C.E. McIlwain
                                    Observations of High Intensity
                                    Radiation by Satellites 1958,
                                    Alpha and Gamma
                                    Jet Propulsion, 28, 588, 1958

L16 Pfotzer, G., A. Ehmert, H. Erbe, E. Keppler, B. Hultqvist,
          and J. Ortner   A Contribution to the Morphology
                                    of X-Ray Bursts in the Auroral
                                    Zone
                                    J. Geophys. Res. 67, 575 - 585,
                                    1962

L17 Winckler, J.R., L. Peterson, R. Arnoldy, and R. Hoffman
        X-Rays from Visible Aurorae at Minneapolis
        Phys. Rev. 110, 1221, 1958

L18 Anderson, K.A.
        Soft Radiation Events at High Altitudes During the Magnetic Storm of August 29-30, 1957
        Phys. Rev. 111, 1397-1405, 1958

L19 Winckler, J.R., L. Peterson, R. Hoffman, and R. Arnoldy
        Auroral X-Rays, Cosmic Rays, and Related Phenomena During the Storm of February 10-11, 1958
        J. Geophys. Res. $\underline{64}$, 597, 1959

L20 Anderson, K.A.
        Balloon Observations of X-Rays in the Auroral Zone I
        J. Geophys. Res. $\underline{65}$, 551-564, 1960

L21 Anderson, K.A. and D.C. Enemark
        Balloon Observations in the Auroral Zone II
        J. Geophys. Res. $\underline{65}$, 3521, 1960

L22 Brown, R.R.
        Balloon Observations of Auroral Zone X-Rays
        J. Geophys. Res. $\underline{66}$, 1379, 1961

L23 Little, C.G. and H. Leinbach
        The Riometer, a Device for the Continuous Measurements of Ionospheric Absorption
        Proc. IRE, 47, 315-320, 1959

L24 Quenby, J.J. and G. J. Wenk
Tables of Cosmic Ray Threshold Rigidities
Imperial College of Science and Technology, London, England,
Cosmic Ray Group Report

L25 Neher, H.V. and A.R. Johnston
Modification to the Automatic Ionization Chamber
Rev. Sci. Instr. $\underline{27}$, 173-174, 1956

L26 Greisen, K.
The Intensities of the Hard and Soft Components of Cosmic Rays as Functions of Altitude and Zenith Angle
Phys. Rev. 61, 212, 1942

L27 Pomerantz, M.A.
The Properties of Cosmic Radiation in the Lower Atmosphere
Phys. Rev. $\underline{75}$, 1721, 1949

L28 Wilson, R.R.
Range, Straggling and Multiple Scattering of Fast Protons
Phys. Rev. $\underline{71}$, 385, 1947

L29 Erbe, H.
A Balloone-Borne Apparatus for Measurements of Ionizing Radiation in High Altitudes
Technical Report No. 1, AFOSR Contract No. AF 61(052)-372,
Max-Planck-Institut für Aeronomie,
Institut für Stratosphären-Physik,
Lindau / Harz, Germany (1961)
published also in Elektronik, Nr.3
März 1962, p. 69 - 72

L30 Ehmert, A., H. Erbe, G. Pfotzer, C.D. Anger, and R.R. Brown
Observations of Solar Flare Radiation and Modulation Effects at Balloon Altitudes, July 1959
J. Geophys. Res. $\underline{65}$, 2685 - 2694, 1960

L31 Keppler, E., A. Ehmert, G. Pfotzer
Solar Proton Injections During the Period from July 12th to July 28th, 1961 at Balloon Altitudes in the Auroral Zone (Kiruna/Sweden)
Space Res. III, ed. by W. Priester
North Holland Publ. Company
(COSPAR Meeting, Washington, May 1962)

L32 Hoffman, D.J. Hoffman and J. R. Winckler
Simultaneous Balloon Observations at Fort Churchill and Minneapolis During the Solar Cosmic Ray Events of July 1961
to be published in Space Research II, ed. by W. Priester, North Holland Publishing Company
(COSPAR Meeting, Washington, May 1962)

L33 Pieper, G.F., A.J. Zmuda, and C.O. Bostrom
Solar Protons and the Magnetic Storm of July 13, 1961
Space Res. III, ed. by W. Priester
North Holland Publ. Company
(COSPAR Meeting, Washington, May 1962)

L34 Keppler, E., A. Ehmert, G. Pfotzer, and J. Ortner
Sudden Increased Radiation Coinciding with a Geomagnetic Storm Sudden Commencement
J. Geophys. Res. (in press)

L35 Brown, R.R., T.R. Hartz, B. Landmark, H. Leinbach, J. Ortner
Large Scale Electron Bombardment of the Atmosphere at the Sudden Commencement of a Geomagnetic Storm
J. Geophys. Res. 66, 1035 - 1041, 1961

L36 Wakatsuki, T., Y. Hirao, E. Okada, J. Miura, K. Sugimoto, A. Mizobuchi
Gamma Rays from Several Elements Bombarded by 10 and 14 MeV Protons
J. Phys. Soc. of Japan, 15, 1141 - 1150, 1960

**Verzeichnis der Mitteilungen aus dem Max-Planck-Institut
für Physik der Stratosphäre**

Nr. 1/1953   Über den Beitrag der von μ-Mesonen angestoßenen Elektronen zu den Ultrastrahlungsschauern unter Blei. G. Pfotzer

Nr. 2/1954   Ein Zählrohrkoinzidenzgerät zur Registrierung der kosmischen Ultrastrahlung. A. Ehmert

Eine einfache Methode zur Einstellung und Fixierung des Expansionsverhältnisses von Nebelkammern. G. Pfotzer

Nr. 3/1954   Optische Interferenzen an dünnen, bei -190°C kondensierten Eisschichten. Erich Regener (vergriffen)

Nr. 4/1955   Über die Messung der Temperatur des atmosphärischen Ozons mit Hilfe der Huggins-Banden. H. Zschörner und H. K. Paetzold

Nr. 5/1956   Ein neuer Ausbruch solarer Ultrastrahlung am 23. Februar 1956. A. Ehmert und G. Pfotzer, vergriffen (erschienen Z. Naturforschung 11a, 322, 1956)

Nr. 6/1956   Das Abklingen der solaren Ultrastrahlung beim Ausbruch am 23. Februar 1956 und die geomagnetischen Einfallsbedingungen. A. Ehmert und G. Pfotzer

Nr. 7/1956   Die Impulsverteilung der solaren Ultrastrahlung in der Abklingphase des Strahlungseinbruches am 23. Februar 1956. G. Pfotzer

Nr. 8/1956   Die atmosphärischen Störungen und ihre Anwendung zur Untersuchung der unteren Ionosphäre. K. Revellio

Nr. 9/1956   Solare Ultrastrahlung als Sonde für das Magnetfeld der Erde in großer Entfernung. G. Pfotzer

*

Die vorstehenden Hefte können beim Max-Planck-Institut für Aeronomie, 3411 Lindau angefordert werden.

Mitteilungen aus dem Max-Planck-Institut für Aeronomie

Nr. 1 (S) Waibel: Messungen von Primärteilchen der kosmischen Strahlung.

Nr. 2 (S) Erbe: Auswirkung der Variationen der primären kosmischen Strahlung auf die Mesonen- und Nukleonenkomponente am Erdboden.

Nr. 3 (I) Kohl: Bewegung der F-Schicht der Ionosphäre bei erdmagnetischen Bai-Störungen.

Nr. 4 (I) Becker: Tables of ordinary and extraordinary refractive indices, group refractive indices and $h'_{o,x}(f)$-curves for standard ionospheric layer models.

Nr. 5 (S) Schröpl: Über eine Neubestimmung des Absorptionskoeffizienten von Ozon im Ultraviolett bei kleinen Konzentrationen.

Nr. 6 (S) Erbe: Ergebnisse der Ballonaufstiege zur Messung der kosmischen Strahlung in Weissenau und Lindau.

Nr. 7 (S) Meyer: Elektromagnetische Induktion eines vertikalen magnetischen Dipols über einem leitenden homogenen Halbraum.

Nr. 8 (I u. S) Dieminger und Mitarb.: Die geophysikalischen Ergebnisse des 12. - 14. November 1960.

Veröffentlichungen in Vorbereitung

(I) Dieminger und Mitarb.: Die Ionosonde des Max-Planck-Instituts für Aeronomie.

(I) Umlauft: Die Absorptionsmeß-Sonde des M.P.I. für Aeronomie.

(I) Schwentek: Druckzählgerät zur laufenden Registrierung halbstündiger Häufigkeitsverteilungen von Feldstärken.

(S) Ehmert u. Revellio: Tafeln zur graphischen Auswertung von Wellenformen mit mehrfach reflektierten Strahlungsimpulsen von Blitzen auf Reflexionshöhe und Blitzentfernung.

(S) Ehmert, Erbe, Pfotzer: Beschreibung der Anlagen des Instituts zur Registrierung der Neutronen und der Mesonen im Geophysikalischen Jahr 1957/58.

If you have any concerns about our products,
you can contact us on
ProductSafety@springernature.com

In case Publisher is established outside the EU,
the EU authorized representative is:
**Springer Nature Customer Service Center GmbH
Europaplatz 3, 69115 Heidelberg, Germany**

Printed by Libri Plureos GmbH
in Hamburg, Germany